Charlie looks into

greater than, less than

& equal to

Sarah Reeves

Sarah's Books Pty Ltd, PO Box 1904 Armidale NSW 2350, Australia

First Edition, 2024
First published, 2024

ISBN: 978-1-7635799-6-5
Independently published.

Printed in Sydney, NSW, Australia, if purchased in Australia.
eBook also available.

A catalogue record for this book is available from the National Library of Australia.

This book was created, written and illustrated in Armidale NSW, Australia, which is Anaiwan country.

Links to the Australian Mathematics Curricula

ACARA Mathematics Curriculum link:
Foundation - AC9MFN06 - represent practical situations that involve equal sharing and grouping with physical and virtual materials and use counting or subitising strategies.

NSW Mathematics K-10 Curriculum link:
Early stage 1 - MAE-FG-02 - forms equal groups by sharing and counting collections of objects.

Victorian Mathematics F - 6 Curriculum link:
Foundation - VC2MFN01 - name, represent and order numbers, including zero to at least 20, using physical and virtual materials and numerals.

Links to curricula:
The Australian Curriculum, Assessment and Reporting Authority (ACARA) Mathematics F-10 Version 9.0
The NSW Education Standards Authority (NESA) Mathematics K-10 Curriculum (February 2024)
The Victorian Curriculum and Assessment Authority Mathematics Foundation to Level 6 V 2.0

This book is dedicated to Libby.

Thank you.
Thank you for all your help and guidance and
always being just a phone call away.

Charlie looks up and sees

one old Gum tree.

One is greater than zero.

1 > 0

Charlie then sees two rocks.

2 is greater than 1.

2 > 1

Charlie finds two

yellow flowers and

two green capsicums.

2 is equal to 2.

2 = 2

Charlie notices

two orange leaves

on her skirt and

more than ten green leaves.

2 is less than 10.

2 < 10

Charlie plays in the

autumn leaves.

Charlie sees that the number

of leaves on the ground

is greater than

the number leaves in the tree

above her.

In spring time, Charlie looks

up at the same tree

and observes that

the number of branches

is less than

the number of blossoms.

There is one giant Gum tree

and one of Charlie.

There is an equal number

of each.

1 = 1

One number is greater than

another number

if it is larger than it.

In the shadows, Charlie

spots 2 kangaroos and

the trunk of 1 large

Gum tree.

2 is greater than 1.

2 > 1

One number is less than

the other number

if it is smaller.

In Charlie's garden, she finds

3 red tomatoes and

6 yellow tomatoes.

3 is less than 6.

3 < 6

Two numbers are equal

if they are the same.

1 hopping kangaroo and

1 green bush.

1 is equal to 1.

1 = 1

The End.

Other books available in the Charlie series:

Charlie counts to five, on a picnic
ACARA Mathematics Curriculum link AC9MFN02
NSW Mathematics K-10 Curriculum link MAE-RWN-001 &
MAE-RWN-002
Victorian Mathematics F - 6 Curriculum link VC2MFN02

Charlie explores whole number 7
ACARA Mathematics Curriculum link AC9MFN06
NSW Mathematics K-10 Curriculum link MAE-RWN-01
Victorian Mathematics F - 6 Curriculum link VC2MFN06

Charlie loves to share, exploring odd and even numbers
ACARA Mathematics Curriculum link AC9MFN06
NSW Mathematics K-10 Curriculum link MAE-FG-02 & MA1-FG01
Victorian Mathematics F - 6 Curriculum link VC2MFN06 & VC2M3N01

Charlie finds a pattern, in the sandpit
ACARA Mathematics Curriculum link AC9MFA01
NSW Mathematics K-10 Curriculum link MAE-FG-01
Victorian Mathematics F - 6 Curriculum link VC2MFA01

Charlie goes for a walk, introducing algebra, with oranges
ACARA Mathematics Curriculum link AC9MFA01 & AC9M1A02
NSW Mathematics K-10 Curriculum link MAE-FG-02
Victorian Mathematics F - 6 Curriculum link VC2MFA01 & VC2MFA02

Other books available in the Charlie series:

Charlie loves to cook, exploring measurement
ACARA Mathematics Curriculum link AC9MFM01
NSW Mathematics K-10 Curriculum link MAE-CSQ01
Victorian Mathematics F - 6 Curriculum link VC2MFM01

Charlie explores fractions
ACARA Mathematics Curriculum link AC9M2M02
NSW Mathematics K-10 Curriculum link MAE-GM-03
Victorian Mathematics F - 6 Curriculum link VC2M2M02

Charlie finds Geometry, in nature
ACARA Mathematics Curriculum link AC9MFSP01
NSW Mathematics K-10 Curriculum link MAE-2DS-01 & MA1-2DS-01
Victorian Mathematics F - 6 Curriculum link VC2MFSP01

Links to curricula:
The Australian Curriculum, Assessment and Reporting Authority (ACARA)
Mathematics F-10 Version 9.0
The NSW Education Standards Authority (NESA) Mathematics K-10
Curriculum (February 2024)
The Victorian Curriculum and Assessment Authority Mathematics Foundation
to Level 6 V 2.0

www.ingramcontent.com/pod-product-compliance
Lightning Source LLC
Chambersburg PA
CBHW042120060426
42449CB00030B/40